Six-Ton Trucks

A VISUAL HISTORY OF THE ARMY'S MOST VERSATILE HEAVY TRUCK 1941-1950

by David Doyle

Published by
Ampersand Group, Inc.
A HobbyLink Japan company
235 NE 6th Ave., Suite B
Delray Beach, FL 33483-5543
561-266-9686 • 561-26609786 Fax
www.ampersandpubco.com • www.hlj.com

Acknowledgements:

This book would not have been possible without the considerable help of Tom Kailbourn, Jim Gilmore, Pat Stansell, Eric Reinert, John Blackman, Bryce Sunderlin, Sean Hert, Scott Taylor, Brett Stolle, Jim Davis, the late Kevin Kronlund, and of course my wonderful wife Denise.

Sources:

TM 9-813 War Department Technical Manual, 6-ton, 6x6 Truck (White, Corbitt, and Brockway), 19 February 1944.

ORD 8-9, SNL G-512 and SNL G-514, Higher Echelon Spare Parts and Equipment (Addendum) (ORD 8-9) for Truck, 6-ton, 6x6, Prime Mover, w/ or w/o winch (Corbitt Model 50SD6) (SNL G-512) and Truck, 6-ton, 6x6, Prime Mover, w/winch, 2,000 gal. gas tank chassis, van, 1942-43-44 (White Model 666) (SNL G-514), 10 December 1944.

Engineering of Transport Vehicles, Chief of Ordnance-Detroit, 1942-1945.

BR-3, History of the Development of Bridging Equipment, III, Heavy Floating Bridging, Army Service Forces, Corps of Engineers, 18 March 1946.

BR-4, History of the Development of Bridging Equipment, IV, Steel-Treadway Bridging, Army Service Forces, Corps of Engineers, 10 June 1946.

BR-11, History of the Development of Bridging Equipment, IV, Transportation for Bridging, Army Service Forces, Corps of Engineers, 10 August 1947.

ME-12, "History of the Development of Mechanical Equipment XII Fire-fighting equipment, Army Service Forces, Corps of Engineers, 6 September 1946.

Front cover: Photographed at a War and Peace military equipment show in the U.K. was this nicely restored Brockway 6-ton 6x6 crane truck with a Quick-Way crane.(John Blackman)

Title page: This restored Corbitt 6-ton 6x6 prime mover bears replica markings for the 127th Ordnance Maintenance Battalion of the 5th Armored Division and has the open cab with a ring-mounted .50-caliber machine gun. (John Adams-Graf)

Rear cover: A restored 6-ton 6x6 bridge-erector truck has been given replica markings for a vehicle attached to Company C, 55th Armored Engineer Battalion, 10th Armored Division in the European Theater of Operations. This open-cab truck is fitted with a ring mount. Loaded vertically in the body are treadway sections with steel mesh road surfaces. (John Blackman)

Table of Contents

During testing operations, a Corbitt 6-ton 6x6 truck rolls over a ponton bridge. With the tailgate lowered, the spare tire on the left front side of the cargo body is in view. Dual rear tires were employed. Each rear mud flap was fitted with three steps. (Office of History, Headquarters, U.S. Army Corps of Engineers)

Introduction

As the U.S. Army prepared for entry into WWII, which seemed increasingly inevitable, there was considerable emphasis on standardization of motor transport vehicles and the components thereof. While during WWI the U.S. Army had fielded a myriad of types of vehicles from a broad range of manufacturers, that bought with it significant logistical and maintenance problems. These problems would haunt the army for years to come. In 1938, speaking to the meeting of the Society of Automotive Engineers concerning the Army's 10,000-vehicle fleet, Assistant Secretary of War Louis Johnson said, "We have 28 different makes and models of passenger cars and 143 makes and models of trucks. In a major emergency, the maintenance of this number of different vehicles in the theater of operations obviously would be an impossible task."

Learning from this painful lesson, on 1 June 1939 the Quartermaster Corps Technical Committee recommended to the Chief of Staff that the army adopt five weight classes of trucks: 1/2-ton, 1 1/2-ton, 2 1/2-ton, 4-ton, and 7 1/2-ton. All tactical vehicles were to be all-wheel drive. Further, Quartermaster General Louis H. Bash, long an advocate of standardization, recommended that trucks be purchased in large lots, rather than small contracts, and that maximum interchangeability of parts be a goal. Acting Chief of Staff George C. Marshall approved these recommendations shortly thereafter.

These initial five categories, however, were quickly revised, as recounted in the Rifkind report: "The 1/4-ton jeep was introduced in 1940, and the 3/4-ton to replace the 1/2-ton, made its debut in 1941. A 4-5 ton, 5-6-ton and 6-ton chassis were also added to fill the gap between the 4 and 7 1/2-ton groups."

One of the functions of the newly added 6-ton group was to serve as a prime mover for heavy anti-aircraft artillery. During the time frame in question, antiaircraft was a function of the Coast Artillery. Therefore, the request for the heavier truck originated from the Coast Artillery Corps, who desired a prime mover for their 3-inch M3 and 90mm M1 anti-aircraft cannon. The truck soon caught the attention of the Corps of Engineers, who desired not only a heavy prime mover, but also a heavy chassis for other uses.

Prime Mover

The army procured two types of 6-ton 6x6 prime movers. In May 1940 87 Mack model NM trucks were delivered, the first of 199 of the type to be purchased that year. Ultimately, 6,886 of the model NM would leave the Allentown assembly line. While U.S. troops used a few of these, particularly the earliest production, the bulk of the type was supplied as foreign aid.

However, in January 1941 104 Corbitt 50SD6 trucks were delivered to the army. These 6-ton 6x6 trucks were similar to the 13 7-1/2 ton 6x6 trucks the Henderson, North Carolina firm had delivered the year before. The initial group of 104 trucks was joined by 87 more in February and a dozen further examples in March. These 203

trucks constituted what is referred to as the Corbitt Series A trucks.

In May 1941 deliveries of a block of 200 more examples of the 50SD6 began. Known as the Series B trucks, the hood and grill of these vehicles (and their successors) were markedly different from those found on the Series A trucks. Series C truck deliveries began with 61 trucks in October 1941 and concluded with the delivery of 39 more examples the next month.

Early trucks were powered by the 779 cubic inch displacement Hercules model HXC 6-cylinder gasoline engine. This engine gave way in later models to the 855 cubic inch Hercules HXD, also a gasoline engine. Either engine was coupled to a Fuller 4-A-86 four-speed manual transmission and Timken T-77 two-speed transfer case, providing the vehicle with a possible 8 forward gears and two reverse. The trucks were equipped with full air brakes. Unlike today's trucks, which feature spring safety brakes, wherein large springs apply the brakes when there is no compressed air, the WWII 6-ton, built before such a system was devised, required compressed air for brake operation. No compressed air—no brakes.

The Corbitt trucks, whose series ultimately ran A through G, were assigned Standard Nomenclature Listing (SNL) number G-512.

However, there was a gap in Corbitt production of 6-ton prime movers between November 1941 and March 1942. But, in January 1942 the army accepted 125 6-ton prime movers, and a further 163 in February. White Motor Company of Cleveland produced these trucks, which were near-clones of the 50SD6. White assigned the vehicles the rather ominous model number "666" (6-ton, 6x6), and the Ordnance Department assigned SNL G-514.

Corbitt resumed production with their Model D in March 1942, production of which ran through May 1942, when the last of these 100 trucks were accepted. The Model Ds, like their predecessors, featured cargo beds manufactured by the Galion Allsteel Body Company. As the name suggests, these bodies were steel, which by early 1942 was becoming in increasingly short supply. The army began ef-

Corbitt 6-ton 6x6 truck U.S.A. number W-51123 negotiates a ponton bridge in November 1941, drawing a 4x4 low-bed transporter trailer with a heavy-duty drawbar and folding boarding ramps at the rear. This type vehicle initially was intended as an antiaircraft artillery prime mover. The earliest examples, such as this one, were known as Series A, and had a notably different hood, radiator and brush guard that was used on subsequent models. (Office of History, Headquarters, U.S. Army Corps of Engineers)

forts to replace steel cargo bodies on all transport vehicles with wooden equivalents. This was the case for the Corbitt 50SD6 and White 666 as well.

The other very noticeable change was the transition from closed cab construction to open cab construction. Open cabs had the advantage of using less steel than their close-cab counterparts, but perhaps even more notably, required less precious cargo space when being shipped overseas. A 29 December 1942 memorandum from Captain R. O. Chapman, Chief, Transport Vehicle Section, to Colonel Edwin Van Deusen, states regarding this "White now plans to start delivery of vehicles with open cabs by March 15th."

The same memo later elaborates "Corbitt may possibly buy open cabs from White and will then be in production on open cab vehicles by April 1st. However, they have also under consideration the building of their required open cabs within their own plant. By their building of their own open cabs, production could start between January 1st and March 1st."

A subsequent memo from 18 March 1943 pushed the initial date for White open cab production to 15 April, with Corbitt anticipating the same date. The October 1944 Chief of Ordnance report "Design, Development, Engineering and Production of Trucks and Semi-trailers" states that the change to open cab 6-tons became "effective in production April and May 1943."

With an open military cab and wooden cargo box, Corbitt kept the 50SD6 in monthly production through June 1945, totaling 3,077 trucks. White production of the prime mover was more sporadic, with July, August and November 1942, as well as February through June 1944 being skipped entirely. Nevertheless, White produced 3,547 examples of the prime mover.

The K-56 Radar Van

The U.S. Army Signal Corps utilized its own designation system for vehicles and trailers into the 1950s. With respect to the 6-ton 6x6, the K-56 designation was assigned to a truck equipped with a van body rather than a cargo box. The van body, which was manufactured by either the Pearly A. Thomas Car Works or Superior Coach Corporation, housed a generator. That generator was used to supply power to first the SCR-268, and later the SCR-545, radar systems which were housed in a full trailer, for which the K-56 doubled as a prime mover. These systems were the gun-laying radar used to control 90-mm antiaircraft guns.

In order to avoid premature obsolescence of closed cab components, the decision was made to use up those parts already in production on the van trucks, whose shipping space inherently could not be significantly reduced by use of open cabs. This detail was recorded in a 18 March 1943 memo from Captain F. R. Nail, Chief Manufacturing Planning Section, to Colonel E. S. Van Deusen, Concerning White's adoption of the open cab, Captain Nail wrote "Closed cab material will be used up on chassis for Signal Corps van trucks."

White produced the K-56 exclusively beginning in March 1942. The bulk of the production of 1,870 K-56 had been completed by September 1943, although the single final truck of the group was accepted in December 1943. However, a letter from the Ordnance Department to the Commanding General, Chief of Ordnance,

Detroit, on 20 March 1945 concerning the disposition of excess vehicles reported by the Chief Signal Officer provides additional insight. Concerning these trucks, the letter reads "Headquarters, ASF, has informed this office that 121 Trucks, 6-ton, 6x6, K56 Van, and 360 Trucks, 2-1/2 Ton, 6x6, K60 Van, are declared surplus by Signal Corps at Camp Stewart and will be moved to Holabird for removal of Signal equipment and release to Ordnance as soon as transportation can be arranged."

Following up on this, the next day Lt. Col. William Brown wrote to the Chief of Ordnance concerning these trucks, "Instructions have previously been issued to your office concerning removal of van bodies and utilization of chassis, 6-ton, 6 x 6, and 2-1/2-ton, 6 x 6, and semi-trailers, K-67. In this connection, it will be necessary to plan procurement of winches for conversion of the 120 trucks, 6-ton, 6x6, to prime movers."

At the time this was being discussed, overseas supply routes of the allied armies were lengthy, and the supply chain was stretched to the limit. What the U.S. Army termed "heavy-heavy trucks" – those over 4-ton capacity – were desperately needed and in extremely short supply. This situation was reflected in a letter from Brigadier General F. A. Heileman, Director of Supply, concerning these trucks. General Heileman wrote "Of the above vehicles trucks, 6-ton, 6 x 6, K-56 and 2-1/2-ton, 6x6, K-60, are constructed on chassis which are currently in critical supply status. The 6-ton, 6x6, chassis is readily convertible to the truck, 6-ton, 6x6, prime mover; the 2-1/2-ton, 6x6, chassis can be converted to Ordnance cargo trucks or supplied to other Technical Services."

General Heileman advocated either driving the trucks from Camp Stewart, Georgia, to Fort Holabird, Maryland, or stripping the trucks at Camp Stewart and turning them over to Ordnance at that location, saying "The importance of prompt transfer of the above equipment cannot be stressed too strongly."

Bridge Erector

The concept of the bridge erector truck dates to June 1941, when the Engineer Board was directed to develop "a treadway-handling device for mounting on the trucks used for transporting the treadways themselves. The initial such vehicle was a 6-ton 4x4 made by the Four Wheel Drive Auto Company (FWD) of Clintonville, Wisconsin. The truck featured an air compressor and air storage tanks for inflating floats, and a winch and crane for handling treadways. After successful tests of a pilot, a further 28 examples were purchased on contract W-145-ENG-249, dated 21 August 1941. The crane on these vehicles was in actuality a demountable A-frame tripod.

The 16th Armored Engineer Battalion, who utilized 14 of these vehicles during the Carolina Maneuvers, found the system lacking, owing largely to the amount of time (15 minutes) involved in setting up the A-frame. Lieutenant W. E. Cowley proposed instead a crane consisting of two parallel arms joined rigidly to the rear of the truck, such that they could pivot. The mechanism would be powered by a bullwheel driven from the truck power take-off. Lt. Colonel T. H. Stanley, CO of the 16th, endorsed this proposal, stating it could position treadways in 90 seconds, versus the substantial time of the existing truck, plus having the added benefit of requiring fewer personnel and being safer.

Cowley's design, with Stanley's support, was recommended for further investigation and a request was made for production of an experimental model, funds permitting. This problem was referred to Four Wheel Drive, along with a request for an estimate. On 9 December 1941, FWD submitted a drawing, indicating that the overall width of the vehicle would be at least 110 inches.

In January 1942, as a result of a conference between Captain F. J. Bogardue of the Engineer Board and Colonel Daniel Noce, Engineer of the Armored Force, the decision was made to modify Cowley's design to use hydraulic hoists rather than a bull wheel, in order to provide smoother operation and better control. All agreed that development and production of bridge erector trucks should be expedited.

On 8 January 1942, the Quartermaster Corps, who at that time were responsible for Army motor vehicle procurement, issued contract W398-QM-11751 to the Brockway Motor Company for 655 trucks equipped with the A-frame lifting device.

At about the same time, the Engineer Board instructed the Heil Company of Milwaukee to submit an example of the Armored Force device, the U-frame derrick. This body, as well as the first of the Brockway trucks equipped with an A-frame, arrived at Fort Belvoir on 26 March. After inspection by the Engineer Board, the units were shipped to Fort Knox for testing by the Armored Force.

At a meeting 2-4 April, the Armored Force opined that the A-frame was unsatisfactory, but that the U-frame was useable, although not fully developed. Given the urgent need, the Armored Force stated that a limited number of trucks with the U-frame would be accepted until more satisfactory equipment could be procured.

The Engineer Board recommended to the Chief of Engineers that the first 235 trucks on contract W398-QM-11751 be equipped with the U-frame equipment, and that the Quartermaster General should be advised not later than 21 April if that equipment was to be used on the next 100 trucks, and such a procedure be used thereafter until a final decision on treadway launching equipment was reached.

The Engineer Board meanwhile was investigating other types of treadway handling equipment. These included an alternate design offered by Heil, a FWD-produced, Cowley-designed bullwheel driven model, and a third type, developed by the Daybrook Hydraulic Corporation. The Daybrook offering was similar to the Cowley design, but operated hydraulically rather than through a bullwheel.

The three new types, along with the earlier A and U-frame models, were demonstrated for comparison to representatives of the Armored Force, Engineer Board, Office of the Quartermaster General and Office, Chief of Engineers. This demonstration occurred 29 May 1942 at Fort Knox. At this demonstration, the Daybrook product clearly excelled.

After the demonstration, Colonel Stanley of the Armored Force recommended that the Heil contract for 235 U-frame devices be cancelled, that the Daybrook-designed device be procured, and that these bodies and derricks be procured by the Corps of Engineers and mounted on chassis provided by the Quartermaster Corps.

On 8 June 1942, the Engineer Board issued contract W145-ENG-427 to Daybrook Hydraulic for 308 of their bed, with a contract value of $789,240.50. A further 354 were ordered from Heil on 22 June 1942, on contract W145-ENG-435, valued at $871,833.

The initial model produced was designated M-11, later superseded by the M-11A. Brockway built the initial trucks, and in fact, most of the bridge erector chassis, totaling 1,166 trucks, between July 1942 and March 1944. To augment these vehicles, a further 1,146 bridge erector chassis were built by White from January 1944.

Even these numbers were inadequate. Beginning in September 1944 Ward La France began producing the chassis as well, turning out 489 of 533 contracted units before production of the type was cancelled in June 1945. Four Wheel Drive reentered bridge erector production that month (June), producing 40 chassis of 168 ordered before that contract too was cancelled.

Crane Carrier

The 6-ton crane-carrier chassis was built exclusively by Brockway, beginning in July 1943 and continuing through June 1945. During this time 1,224 units were built. These trucks had a 197-inch wheelbase and rode on 12.00-20 tires. A 25,000-pound capacity double drum winch was mounted at the front. The frame was of especially robust construction. On 6 March 1944, authorization was given to utilize walking beam suspension in crane carrier production rather than standard springs and spring jacks. The trucks, which were given the Brockway designation C666, featured a unique half-cab, the portion to the right of the driver being occupied by the crane boom when configured for transport, precluding passenger seating.

Brockway built 1,224 of the crane carrier chassis between July 1943 and June 1945. A further 809 were cancelled as the war situation improved. Similarly, the 867 examples on order from Four Wheel Drive (FWD) were cancelled before they were produced.

Various models of cranes were mounted on the chassis, most commonly the Quick-Way model E; the Schield-Bantam model M-49A; or Osgood. The Quick-Way was the version used during WWII, and the crane proper was the same as that mounted on the Coleman G-55A 5-ton 4x4. Powering the crane operations was an International-Harvester model Q-W/U—9, 4-cylinder 35 horsepower engine. The crane could also be equipped with pile driver or dipper, clamshell or dragline buckets, making it a very versatile machine indeed. These various accessories were transported on a Timpte mode QW-T-8 trailer, towed by the crane carrier.

The crane had a rated capacity of 13,000 pounds with a 10-foot reach, decreasing to 5,500 pounds at its maximum 25-foot radius. The truck chassis and crane combination tipped the scales at 33,100 pounds.

Tanker

Providing fuel to mechanized forces is a formidable task, and as allied supply lines lengthened during WWII increasing efforts were placed on providing viable bulk transport. Early in 1942 the Quartermaster Corps began developing a 2,000 gallon tank truck based on the 6-ton chassis for use in North Africa. Toward this end, one 2,000-gallon, 4-compartment tank was purchased from the Butler Manufacturing Company on contract W-398-QM-10793. This was mounted on one of the White 6-ton 6x6 chassis purchased on contract W-398-QM-10326. To facilitate desert operation, the vehicle was initially equipped with 10.00-22 tires. After a cross-country drive from Kansas City to Aberdeen, the vehicle was subjected to cross-country tests, during which the frame cracked. This was repaired and the design modified to prevent recurrence. The truck was then shipped to Camp Seeley, California, for desert testing, and the tires changed to size 14.00-20. Testing at Camp Seeley continued for eight months. Although the vehicle was found "extremely satisfactory," Army Ground Forces contended that there was no requirement for such a special purpose tactical vehicle utilizing the 6-ton chassis, which was critically needed as an anti-aircraft gun prime mover. Accordingly, on 12 November 1942 Ordnance Committee action 19180 disapproved standardization of the vehicle.

By late 1943 Army Ground Forces position began to change, and the subject of a fuel tanker utilizing a 6-ton chassis was revisited. Based on reports from the field, it was felt that a truck and tank trailer combination, each with a capacity of 2,500 gallons, should be developed. The Ordnance Committee, by action OCM 22505, made such a recommendation, and development was subsequently approved by OCM 22865 on 10 February 1944.

Whereas the earlier 2,000-gallon tanker utilized the prime mover chassis, the new effort would be built around the more robust Brockway crane carrier chassis. However, rather than utilizing walking beam suspension, then standard on the crane carrier, the standard spring suspension of the bridge erector would be used. Also taken over from the bridge erector was the standard cab, in lieu of the crane carrier half-cab.

On 25 February 1944 Ordnance entered a contract with the Heil Company of Milwaukee for the development of two pilots to be delivered on or about 1 March 1944. The specifications included the crane carrier chassis equipped with a 2,500 gallon tank, and a 2,500 gallon tank trailer, having hubs, brakes, wheels, tires and tank interchangeable with those components of the truck. The truck was designated T-29, and the trailer T-52. Despite the ambitious schedule, the vehicles were not delivered to Ordnance until 4 April 1944.

Pilot number 1 was shipped to the Quartermaster Board, Camp Lee, Virginia, and Pilot number 2 was shipped to the Ordnance Tire Test Fleet, Normoyle, Texas. Pilot 1 was subjected to operational tests, and pilot 2 to durability tests. Quartermaster Board report T-339, 30 July 1944, reported that the vehicles were satisfactory, while the Normoyle operation found some deficiencies. Heil corrected the deficiencies in the tank, but much of the concern expressed by the Normoyle evaluators had to do with the engine and transmission, which they felt was underpowered and did not provide a sufficient number of gear ratios. Those objections were waived by the Quartermasters, as the chassis and power train provided were the only standard units available at that time suitable for the application.

The Quartermaster General requested standardization in August 1944. The combination was classified as Standard by OCM 26328 and 27004, 11 January and 15 March 1945, respectively. The truck was designated M34 and the trailer M27. An ordered was placed for 137 vehicles, with production to begin in October 1945 and run through February 1946, but the conclusion of the war resulted in the cancellation of this order.

Fire Truck

As the size of military aircraft, along with their fuel and bomb loads, increased so did the size and capabilities of air field crash trucks.

In late 1943 the Engineer Board was instructed to develop a new type of truck to meet the demands inherent with the larger aircraft. The Board responded with a development plan for Project ME 499, Truck, Fire, Crash, Class 140, on 8 May 1944. Unfortunately, this project became ensnared in rivalry between the Army Service Forces and Army Air Forces concerning responsibility for the development such equipment. Hence, the project never advance beyond basic inquiries of manufacturers.

Whereas conventional fire trucks carry modest amounts of extinguishing agent (often water) onboard, with bulk supplies being drawn from ponds, streams or hydrants, the fast moving and volatile nature of aircraft fires requires crash trucks bring with them the fire suppressant. This means typically crash trucks are much larger than conventional fire trucks. That agent could be water, foam, or carbon dioxide.

The Truck, Fire, Powered, Crash, Class 155, 6x6, High Pressure Fog Foam was the largest and most capable crash truck of its day. The apparatus beds were produced by both Mack and American LaFrance-Foamite Corporation. Both firms had long histories of producing fire-fighting equipment. Testing of the Class 155 with American LaFrance-Foamite apparatus bed began in June 1944. The first Class 155 with Mack-produced apparatus bed was not received for testing by the Engineer Board until late 1944.

The first 200 examples built were constructed on Kenworth 572 chassis, similar to that found under the M1A1 wrecker, and utilized the Mack-produced fire apparatus beds. Subsequent Class 155 trucks were mounted on Brockway 6-ton 6x6 chassis. Ordnance records indicate that 242 of these chassis were built by Brockway from March 1944 through June 1945. Documentation has surfaced that indicates that 144 of these featured American LaFrance apparatus beds, and 91 were equipped with Mack apparatus beds. The details of the 7-truck discrepancy are unknown.

The Mack-built bodies utilized a Hale ZEY, 325 gallons per minute (GPM), 600 psi high pressure pump powered by a Continental model R-602 6-cylinder engine of 602 cubic inch displacement. The American LaFrance-bodied trucks used the company's own model 155 two-stage centrifugal pump rated at 250 GPM and 600 psi. The pump was powered by an American LaFrance 12-cylinder engine.

Both body types featured a 1,000 gallon water tank and two manually operated high pressure turret nozzles on top of the bed. Mounted above the rear wheels of the truck on either side was a 100-foot $\frac{1}{2}$-inch high pressure hose with hand-held fog nozzles. A third such line was stowed in a compartment behind the cab. Two seat spray nozzles were permanently mounted on either side of the windshield, which served to protect the driver with a wall of water should wind push flames toward the truck.

On 30 January 1945 War Department Circular 36 transferred responsibility for crash trucks to the Army Air Forces.

Truck Tractor

Flotation in soft terrain such as sand is of great concern to vehicle designers, as well as armies operating in the field. The heavy burden placed on engineer equipment, often operating in notably primitive conditions, made this of particular interest to the Engineers. Based on positive initial tests, on 1 June 1944 OCM 24027 approved a re-

quest from the Chief of Engineers to the Chief of Ordnance for development of a kit to convert the 6-ton chassis to 14.00-20 tires. There was initial concern about the increased bearing loads inherent with such a change, but the Chief of Engineers on 11 May 1944 accepted increased maintenance in order to obtain the desired floatation and performance. Ordnance Development Project KG-453 was initiated to achieve this goal.

In April 1944 the Engineers were experimenting with a truck-tractor created by mounting a fifth wheel in lieu of the cargo bed on a 6-ton prime mover. The tractor's intended purpose was to tow the Corps of Engineers 20-ton low bed semitrailer. This configuration resulted in the rear suspension of the truck being so overloaded that the axle shafts broke and the gears in the transfer case were stripped during off-road operations.

Seeking to remedy this, the Development Division, Office, Chief of Ordnance-Detroit, (OCO-D) conferred with the Engineer Liaison Officer (OCO-D), seeking an immediate means of overcoming deficiencies with the axles of 6-ton trucks. As a result of this, on 26 May 1944 the Chief of Engineers requested that the Chief of Ordnance develop a rear chain drive bogie for the standard 6-ton 6x6, suitable for installation of 14.00-20 dual tires. Such a development was recommended by OCM 24369 on 24 June 1944, and approved by OCM 24586 on 13 July 1944.

Two pilot vehicles were authorized for procurement under this project. Contract W-04-200-ORD-543, RAD 2856 was placed with the Sterling Motor Truck Company of Milwaukee, Wisconsin for the installation of such a bogie on a standard 6-ton 6x6. This truck would be designated the T27E1. A similar contract, W-04-200-ORD-543, RAD 2873 was placed with the Los Angeles firm of Cook Brothers, for the second pilot, designated T27E2.

The vehicles differed in the design of the chain-drive jackshaft, with Sterling using three Krohn differentials that claimed to deliver at least 25% of the available torque to any one wheel, regardless of traction. Sterling dubbed the set-up the 265 WX Super-traction. Cook Brothers opted for a conventional differential. Both types proved to be acceptable, as recorded in Engineer Board Report 925 of 20 April 1945, but neither was adopted and work on the units was terminated on 25 August 1945.

As an alternative, OCM 24696 of 10 August 1944 authorized development by Timken-Detroit Axle Company of gear drive rear axle housings for the 6-ton that would properly accommodate 14.00-20 dual tires.

The conventional-drive Truck, 6-ton, 6x6, Tractor was standardized by OCM Item 27191 in April 1945. Very similar to the basic prime mover, the truck had a 185-inch wheelbase and featured a low-mounted heavy-duty fifth wheel. Either 14.00-20 or 12.00-20 tires could be used. White was the sole producer of these vehicles, completing 112 from May through August 1945.

Prime Mover

Between the cab and the body was the mid-vehicle-mounted winch, employed in emplacing artillery and in self-recovering the vehicle should it become mired. The winch was chain-operated from the power take-off on the transfer case. The winch cable was routed to the rear of the chassis by means of sheaves. Also shown is the method of lashing the tarpaulin to the cargo body. (Military History Institute)

266-3-41 H.Q.D.
CORBITT 6 T. 6x6

Extending on a drive shaft to the right side of the winch was a cathead, also called a cap-stan: a small drum around which a line could be wrapped to be powered by the winch. The shaft for the cathead was attached to the worm-drive gear case of the winch. The U.S.A. number on the hood of this Corbitt 50SD6 is W-51656. (ATHS)

In a head-on view of a Corbitt 6-ton 6x6 truck, the forward pintle hook is clearly visible below the bow in the front bumper. Faintly visible above the grill is a nameplate that reads "CORBITT / HENDERSONVILLE, N.C." A very close inspection of the photo reveals that the crosspieces on the grill were rods with a noticeable twist to them. The vertical pieces of the grill were metal strapping. Also visible is the Timken F-3100-W-X-6 front axle. (ATHS)

A step was attached to each side of the top edge of the tailgate, and three steps were fastened to each rear mud flap. Below the center of the tailgate was a toolbox door, hinged at the bottom, secured with a padlock. Between the bumperettes was the rear pintle hook. The rear axles were the Timken SD-353-W-X-7 model. (ATHS)

On display at a U.S. Army Quartermaster Corps exhibit is a Corbitt 6-ton 6x6 truck, U.S. Army registration number W-51656. Whereas the grills of these vehicles were made of a mix of twisted rods and metal strapping, the headlight brush guards were made of untwisted rods with a metal-strapping frame. (NARA)

For antiaircraft defense, an M49 ring mount was experimentally installed on Corbitt 6-ton 6x6 truck U.S.A. number 520401 at Aberdeen Proving Ground in January 1944, as part of Project No. 6-11-34-5. The ring mount was installed on three stanchions. A raised hatch was installed on the cab roof to accommodate the gunner. (NARA) **Inset:** The M49 ring mount on U.S.A. number 520401 is seen from the left rear, showing the manner in which the left rear stanchion was installed on two special brackets: one at the middle of the stanchion and one at the bottom. Holders for two ammunition boxes were affixed to the ring. (NARA)

In a driver's view of the cab of a Corbitt 6-ton 6x6 truck, on the left side of the steering column is the control for trailer electric brakes. Below the dashboard is a fire extinguisher. Above the accelerator pedal is the starter switch. The tall lever to the right is the transmission gearshift lever. (ATHS)

White Motor Company built 3,547 6-ton 6x6 prime-mover trucks from January 1942 to June 1945, designating these vehicles the Model 666. It was quite similar in shape and design to the Corbitt Model 50SD6, from the shapes of the front clip, screened vents in the hood, the front bumper and pintle hook, and brush guards, to the steel cargo body and mid-mounted winch. (ATHS)

"U.S.A." has been painted on the hood of this White Model 666, but the U.S. Army registration number has yet to be applied. Above the top center of the grill is a White manufacturer's plate. On the cab roof is a ventilator hood. (Military History Institute)

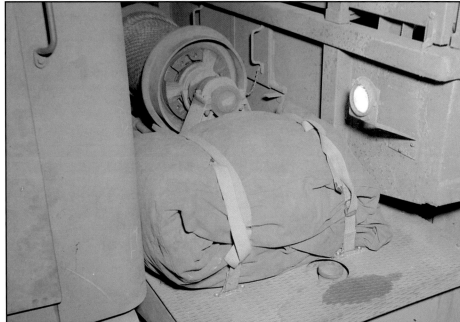

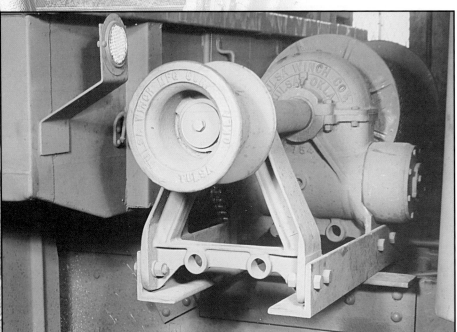

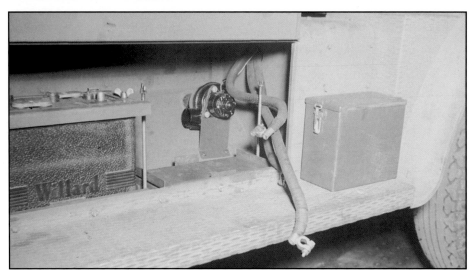

Top left: In a photo of the winch of a White 6-ton 6x6 prime mover from the right side, the cathead is to the right, with the worm-gear casing and the cable drum to the left. The winch drive chain from the power take-off is visible. On the sheave bar, the cross-shaft mounted to the rear of the winch assembly, is a sheave that acted as a cable guide. Below the cathead are the muffler and the tailpipe. **Top right:** The fuel tank and the winch are seen from the left side of a White 6-ton 6x6 prime mover, showing the diamond-tread platform atop the tank. A cutout in the center of the platform provides clearance for the fuel filler and cap. The tank had a capacity of 80 gallons. A tarpaulin is strapped to footman loops on the platform. **Above left:** Cast into both the worm-gear casing and the outer edge of the cathead is the manufacturer's name, the Tulsa Winch Company, of Tulsa, Oklahoma. To the rear of the cathead is a reflector on a sheet-metal bracket. **Above right:** The battery-box door of a White 666 is open, showing one Willard battery in place. There were two such batteries, and the forward one has been removed. The batteries were secured in place with a hold-down bracket that fit over rods equipped with wing nuts. (ATHS, all)

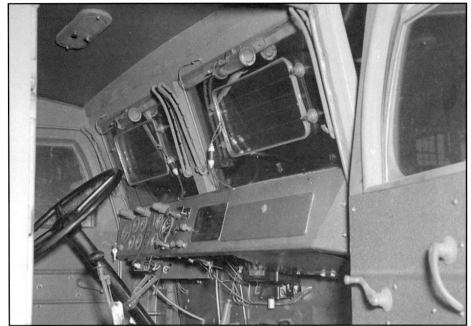

Top left: Electric-powered defrosters have been installed on the rear sides of the windshield of this White Model 666. Above the defrosters are air-powered windshield-wiper motors. On the ceiling of the cab is a small door for operating the roof ventilator. **Above left:** The layout of the instruments and controls on the dashboard of this White 6-ton 6x6 truck differs in several particulars from that of the Corbitt 6-ton 6x6 shown previously. At the bottom of the dashboard above the steering column is a control panel for the windshield defrosters. **Right:** With the right side of the hood of a White 666 open, part of the engine compartment is in view, including the oil-bath air cleaner and the dual horns that are mounted on the firewall. The Zenith carburetor and the carburetor air-intake hose are not mounted in this photo. The two studs on the intake manifold that secure the carburetor in place are below the fronts of the horns. (ATHS, all)

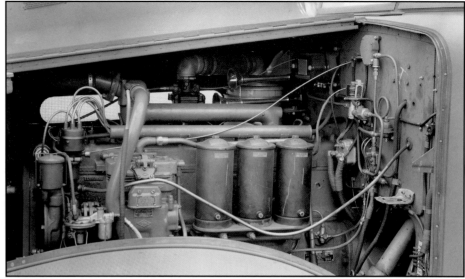

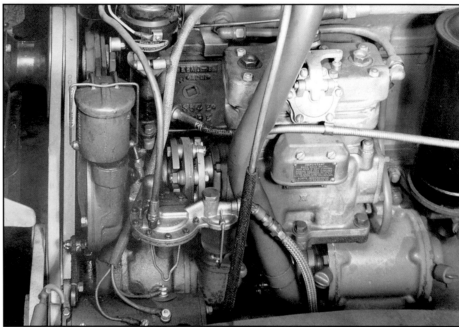

Top left: The air-intake hose and the carburetor are installed on this Hercules engine. The manufacturer's name is in raised letters on the side of the intake manifold. The label on the air cleaner identifies it as an Air-Maze brand. Below the carburetor, at the center of the exhaust manifold, is the heat-control valve, with a "heat on" and "heat off" switch. (ATHS)

Top right: Dominating the left side of the engine are the triple oil filters. Above the filters, and lower than the water connecting tube, is the ignition wiring conduit tube. Toward the left is the air compressor, and to the far left is the water bypass tube. (Bryce Sunderlin collection) **Above left:** As seen from the left side of the engine compartment, the black object

at the center of the firewall is the voltage regulator. The fixture toward the upper right is the pressure regulator, below which is the horn valve. To the bottom right is the lower part of the steering column. **Above right:** In a view of the forward part of the left side of a 1941 Hercules engine as installed in a White Model 666, to the far left are the fan belts, next to which are, top to bottom, the ignition coil, the distributor, the oil-filler spout, the fuel pump, and the generator. Toward the right is the air compressor and, below it, the water pump. (ATHS, both)

With a 6-ton 6x6 prime mover on a rack, many features of the underside of the chassis and front suspension are in view, including the support structure for the front pintle hook. Fastened to the front cross member of the chassis frame are brackets for a service-air and an emergency-air hose and coupling. The U-bolts of the springs are routed through holes in flanges built into the axles. (ATHS)

Left: A completed or nearly completed White Model 666 truck is on the assembly line next to a few half-tracks. The lashings for the tarpaulins, when new, were of a very bright appearance. Rear-view mirrors were mounted on each side of the vehicle. **Right:** During the disassembly of a White Model 666 truck, a process called the tear-down, the cab is being hoisted free of the chassis. The front clip, cargo body, fuel tank, and tandem wheels already have been removed. A clear view is provided of the steering gear. (ATHS, both)

White Model 666 trucks were completely built before being partially disassembled for packing for shipment overseas. Here, the steel cargo body of a Model 666 is being hoisted from the chassis in preparation for packing. The cables are attached to tarpaulin-lashing hooks. (Bryce Sunderlin collection)

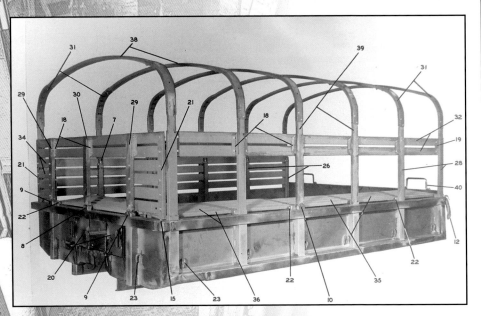

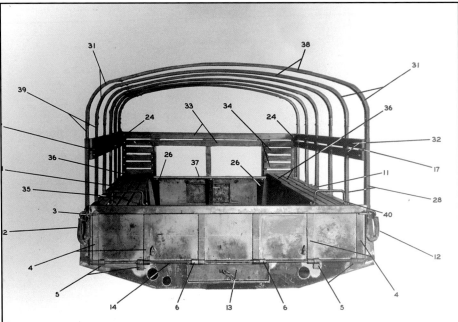

Top left: The early-type steel cargo body of a 6-ton 6x6 prime mover is seen from the front left. The troop seats on the near side are lowered, and the troop seats on the far side are raised. The wooden bows and stakes were of one unit, with metal reinforcing at the bends. The lazy backs were fastened to the stakes with brackets. **Top right:** The steel cargo body is viewed from the rear; the two spare-tire carriers in the front corners of the body have been collapsed, and the troop seats extend from front to back of the body. Below the tailgate is the door to the built-in toolbox. (ATHS, both) **Above:** During the packing of a White Model 666 prime mover sometime in 1942, the axles and winch already have been loaded into the bottom of the shipping crate, and the chassis is being lowered in place. Fragile components such as the carburetor, radiator, fan, steering gear and column, and shifting levers have been removed and packed separately. (Bryce Sunderlin collection)

Top: A round opening was cut in the cab roof of U.S.A. number 55513 and a raised hatch with a rolled upper edge was welded in place. The gunner stood on the seat in order to fire the machine gun. The stanchion brackets were fastened to the bodywork with large hex screws. **Above:** In October 1943 as part of Project 6-11-34-5, the Ordnance Department conducted tests of a ring mount for a .50-caliber machine gun on a White Model 666 prime mover at Aberdeen Proving Ground. The installation of the ring mount was similar to that on the Corbitt 6-ton 6x6 prime mover illustrated earlier and also part of Project 6-11-34-5. (NARA, both)

Above: White Model 666 U.S. Army registration number 55513 was the vehicle used for the experimental ring-mounted machine gun in the October 1943 tests at Aberdeen. Attached to the two stanchions for the ring mount that are in view were brackets that were bolted to the cab. The front bracket was a form-fitting plate that was bolted to the top of the cowl, while the rear bracket was a long, vertical piece of channel iron. **Right:** In order to reduce the profile of the 6-ton 6x6 prime mover, an open-cab version was developed. This reduced the space required for shipping the partially disassembled trucks. Also, a wooden cargo body was introduced, replacing the early, all-steel body. Brackets for installing a ring mount were standard items to the front and to the rear of the passenger's door, as well as on the rear of the cab. (NARA, both)

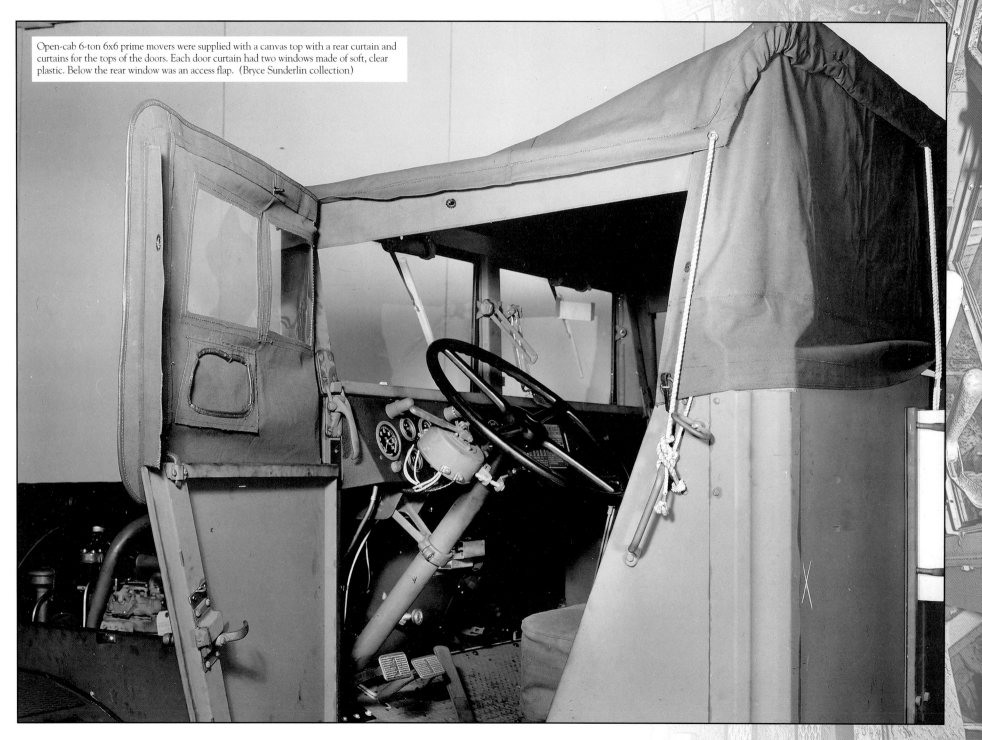

Open-cab 6-ton 6x6 prime movers were supplied with a canvas top with a rear curtain and curtains for the tops of the doors. Each door curtain had two windows made of soft, clear plastic. Below the rear window was an access flap. (Bryce Sunderlin collection)

The door flaps were secured by four fasteners. Attached to the curtain, they comprised leather straps affixed to a metal ring. The leather straps were inserted through loops affixed to the top and to the bottom of the windshield frame and to each top corner of the door, thus holding the curtain in place. (Bryce Sunderlin collection)

In a view of the interior of the open cab of White 6-ton 6x6 prime mover Ordnance serial number 279481 in 1944. The right seat back is formed of diamond-tread steel. When the seat back was lowered, this gave the gunner a non-slip platform to stand on while firing the ring-mounted machine gun. Two universal rifle brackets are at the rear of the cab. Below the dashboard to the far right is a spare-parts container. (TACOM LCMC History Office)

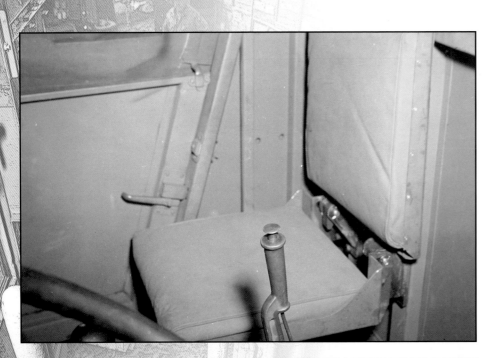

Top left: The right seat in an open-cab White 6-ton 6x6 prime mover is shown with the seat back in the raised position. The interior operating handle for the door and some of the framework of the cab interior are apparent. **Above left:** Here, the hinged seat back has been lowered, now forming a small but adequate platform on which the machine gunner could stand when operating the ring-mounted weapon. The diamond tread is visible. (ATHS, both)

Above right: In a view through the left door of an open-cab 6-ton 6x6 prime mover, the driver's seat, located between the two universal rifle brackets, has been removed. The right seat bottom, which, like the seat back, was hinged, has been raised to provide access to the stowage compartment beneath the seat. This bin was not present in the cab in the preceding photos. (Bryce Sunderlin collection)

Top: The right side of the engine compartment of White 6-ton 6x6 prime mover Ordnance number 279481 is shown in a March 1944 photograph. The air intake hose from the oil-bath air cleaner is routed to the rear of the carburetor instead of to the top, as seen in the earlier photograph of the engine of a 1941 White Model 666 prime mover. **Bottom:** The left side of the engine compartment of White 666 Ordnance number 279481 is displayed in a 3 March 1944 photograph. The triple oil filters are lined up alongside the engine block. Part of the fan is visible to the far left. (TACOM LCMC History Office, both)

Top left: The cab of this 1944 White Model 666 prime-mover truck, U.S.A. number 5114691, was fitted with an arctic top closure during tests at the Detroit Arsenal on 22 July 1948. The truck was originally built as an open-cab version. **Top right:** The arctic top closure is seen from the left side on White Model 666 U.S.A. number 5114691 on 22 July 1948. The closure kit included an insulated hardtop, doors, and door curtains with windows. (TACOM LCMC History Office, both) **Above:** A column of trucks are transporting Coast Artillery antiaircraft personnel and M1 or M1A1 90mm gun mounts during a training exercise in the Mojave Desert in April 1942. The first, third, and fifth vehicles are 6-ton 6x6 prime movers. G.I.s wearing M1917 helmets and wielding rifles are standing in the cargo bodies. (NARA)

Top left: A pair of 6-ton 6x6 prime movers descend a steep grade. Both vehicles have unusual X-shaped devices on the radiator grills as well as equally unusual bins on top of the cab roofs. The fronts of these bins are cut out at the bottom, no doubt to admit air to the roof vents. **Top right:** During U.S. Army tests, two open-cab 6-ton 6x6 prime movers negotiate a gully. The second vehicle sports the X-shaped device on the grill. The bumper of the first vehicle sports the marking, "U.S. ARMY TEST 288." An unusual feature on the second truck is the mounting of the spare tire behind the driver. (Military History Institute, both) **Above:** Members of an antiaircraft crew are towing 3-inch Antiaircraft Gun M3 on Mounting M2A1 with a 6-ton 6x6 prime mover bearing U.S. Army registration number 53687, a during training exercise at the Desert Training Center at Indio, California, on 24 September 1942. (NARA)

A G.I. services the engine of a 6-ton 6x6 prime mover at Champagne Landing, Yukon Territory, Alaska, during a wintry day in 1943 during the construction of the Alcan (Alaskan-Canadian) Highway, a military road linking Alaska Territory with the Lower 48. Can-shaped objects, purpose unknown, are attached to the lower part of the cargo body. (US Army Quartermaster Museum)

An open-cab 6-ton 6x6 prime mover packed with troops and supplies tows a trailer with an M3A3 light tank up a mountainous road. The M3A3 light tank was produced mainly for export, and, judging from the terrain and the uniforms of the personnel, it is possible that this equipment was operated by the Chinese in the China-Burma-India Theater. (NARA)

Based on the M3 medium tank chassis, the tank treadway bridge-layer was equipped with a front A-frame boom and had the mission of unloading and positioning treadway bridge components. In this photo, a tank treadway bridge-layer is unloading treadways from a trailer hitched to a 6-ton 6x6 prime mover at Badonville, France, on 3 March 1945. (NARA)

In the aftermath of World War II, U.S. Army 6-ton 6x6 trucks remained active in war-torn Europe, assisting in the rebuilding effort. This Corbitt Model 50SD6, U.S.A. number 520683, assigned to the 31st Engineer Combat Battalion, draws a trailer holding a damaged streetcar destined for repairs in Vienna, Austria, on 8 October 1945. (NARA)

The same Corbitt 50SD6 prime mover of the 31st Engineer Combat Battalion with a trailer hitched behind holding a damaged streetcar is seen at another location. A light-colored placard that reads "ENGR" is on the grill. Stenciled on the bumper, from its right side to its left, are an insignia, "USFA" (U.S. Forces Austria), a recognition star with a circle around it, and "-VAC- B-21." (NARA)

A 6-ton 6x6 prime mover carries troops of the 255th Engineer Combat Battalion, IV Corps, who have been engaged in building this Bailey bridge over a deep gorge along a road outside of Bologna, Italy. A U.S.A. number is present but not legible on the hood and has been applied in blue-drab paint. (NARA)

A restored Corbitt 6-ton 6x6 prime mover participates in a military vehicles meet. It bears replica markings for the 127th Ordnance Maintenance Battalion of the 5th Armored Division and has the open cab with a ring-mounted .50-caliber machine gun.

USA - 541577 - S

5Δ-127-0   14

The front-mounted pintle hook featured on these trucks is plainly evident in this view, as is the wooden construction of the bed.

Top left: The 666 prime mover used 10.00-22 tires, a relatively uncommon size at the time. Such diversity in tire sizes was one of the reasons for the push for standardization that resulted in the post-war M-series vehicles. Note the safety wire spanning the hub. **Top right:** All the components of the six-ton trucks were much larger and heavier than those used on the more common 2 1/2 ton CCKW. The suspension was semielliptic leaf springs.

Above left: Louvers underneath the fenders provided ventilation for the engine compartment. **Above right:** Additional ventilation for the engine compartment was provided through the side access panels. These panels, along with the hood, folded toward the centerline of the truck when opened.

Top left: The cab half doors were relatively thin, simple affairs, but nevertheless provided some protection from the elements. **Top right:** Unlike most of its contemporaries, rather than a single bucket seat for the driver and a bench seat for the co-driver, the 666 had two bucket seats. **Above left:** A single fuel tank mounted on the driver's side of the truck held the 80 gallons of minimum 70 octane gas that fed the truck. This gave the truck a range of 2-300 miles depending upon load and conditions. **Above right:** Steel mudguards were suspended from the wooden cargo bed. The hardwood cargo beds of these and other U.S. tactical vehicles were frequently produced by furniture manufacturers, and hence the fit and finish was somewhat higher than one may suspect.

Left: The midships winch was intended for use emplacing artillery. The wire rope was fed under the bed of the truck for most rearward pulls. However, a door in the front of the cargo bed allowed the wire rope to be fed into the cargo area as well. **Top right:** The bed of the truck was mounted via an array of brackets, bolts and clamps. **Above right:** Although commonly referred to as wooden beds, these beds did include some steel, such as angle reinforcements, bow sockets, and tie-down points. In addition to conserving steel, the wooden beds were touted as being more readily repaired in the field than their metal counterparts.

This view of a restored Corbitt 6-ton 6x6 provides a clear view of the wooden cargo body. Two steps are on each rear mud flap. The hook at the end of the winch cable is attached to the bottom of the right bumperette. (John Adams-Graf)

Top left: The substantial spring pack that was integral with the truck's 33,000 pound plus gross weight rating. Note the "TIMKEN" name cast into the torque rod mounting bracket.
Right: The 6-ton was imposing, even when viewed from the rear. While the components of the cargo bed have a uniform style with that of the composite bed used on CCKW and 4-ton trucks, everything is much larger. **Above left:** The height of the bed was such that the conventional steps mounted on the top of the tailgate did not provide adequate access. Accordingly, these simple steel steps bolted to the rear mudguards were used to span the difference. Just to the right of the taillight can be seen the trailer electrical connection.

Top left: Beneath the rear tool compartment door is located a pair of rollers to guide the winch wire rope, which plays out between the bumperettes. **Top right:** A capstan on the passenger's side of the winch allowed for forward pulls, as well as rigging for unusual loads. Note the yellow forward-facing reflector on the lower front of the cargo bed. **Above left:** Inside the cavernous cargo area are the spare tire racks, and visible at the sides and front are the troop seats, in the stowed, vertical position. At the very front of the bed is an opening to allow the winch rope to be fed into the bed. **Above right:** Just visible in this photo is the roller chain which drove the winch from the back side. Preventative maintenance was critical—too little lubrication could result in a broken chain, which too much lube would attract dust and dirt, causing rapid wear.

Above: Tulsa cast their name and identification on the end of the cathead. The A-frame-like supports of the winch were steel castings, while the mounting brackets were formed from angle iron. **Top right:** Trailer air brake connections were provided at the front of the truck, as well as the rear. Just inboard of the spring and behind one of these brake connections can be seen the exhaust pipe. **Above right:** These trucks had air brakes. The brake actuating lever and piston is visible here. The brake system relied on air pressure to apply the brakes, unlike today's spring safety air brakes that use air pressure to release the brakes. Hence, the truck could be driven without the air lines connected to the towed load, although stopping from road speed would be difficult.

K-56 Radar Van

A number of White Model 666 prime movers were converted to prime movers for mobile antiaircraft radar equipment. Designated the K-56 trucks, the vehicles had a van body furnished by either the P. A. Thomas Car Works or the Superior Coach Corporation. The K-56 contained the power plant for the radar and it pulled a trailer containing the radar equipment. This example was configured for transporting a SCR-268 long-range radar set for the 586th Signal Detachment at Portland Roads, Queensland, Australia, in late March 1943. (NARA)

A trailer containing SCR-268B radar set serial number 416 is hitched to a White K-56 truck. The truck, trailer, and radar equipment were assigned to the 202nd Signal Depot Company at Townsville, Australia, in late April 1943. The girder-type structure protruding from the rear of the trailer was an arm for the radar set; the arm, when erected, held the radar dipoles. (NARA)

These three GMC CCKWs with communications van bodies and the White K-56 to the right contained the complete power unit for SCR-268 serial number 216 in the traveling position. The vehicles were with the 202nd Signal Depot Company at Townsville, Australia, late April 1943. (NARA)

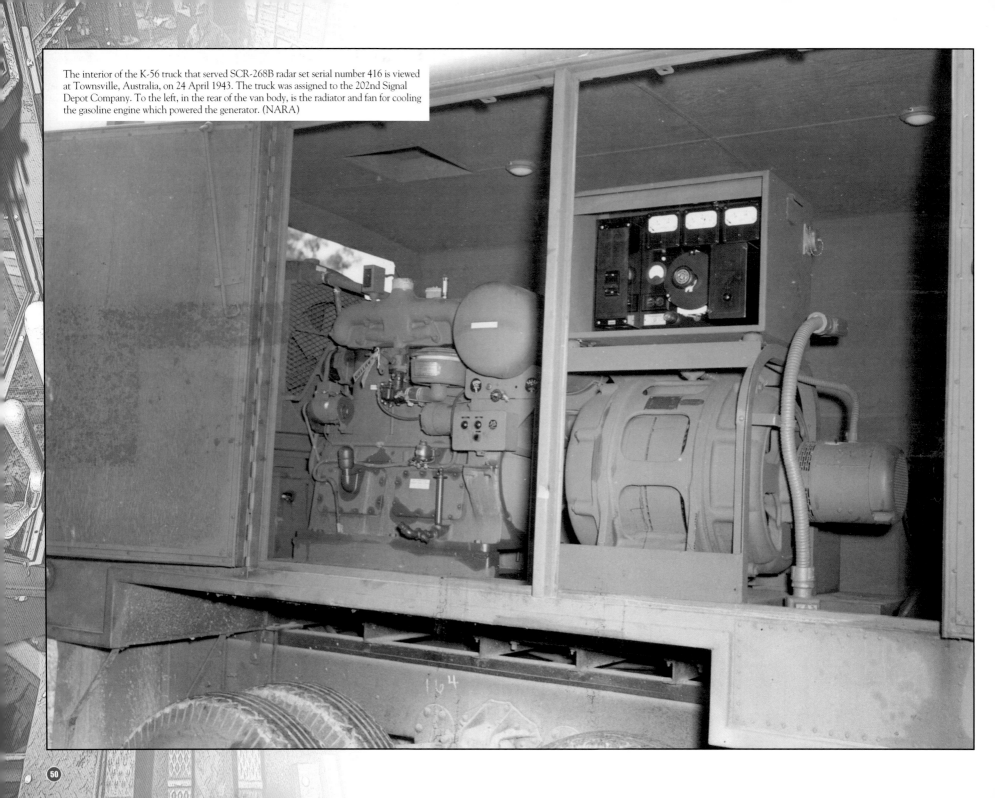

The interior of the K-56 truck that served SCR-268B radar set serial number 416 is viewed at Townsville, Australia, on 24 April 1943. The truck was assigned to the 202nd Signal Depot Company. To the left, in the rear of the van body, is the radiator and fan for cooling the gasoline engine which powered the generator. (NARA)

A K-56 truck, U.S. Army registration number 6021422, is in operation at a radar station on Rendova in the Solomon Islands on 15 July 1943. The power plant in the van body of the vehicle was supplying electricity to the SCR-268 radar set. (NARA)

The same K-56 truck, U.S.A. number 6021422, is viewed from a closer perspective on Rendova. With the side doors open, the generating equipment inside is visible. The generator delivered single-phase, 120-volt, 60-cycle power to the radar set, whether it was the SCR-268, as in this example, or the SCR-545. (NARA)

Bridge Erector

Brockway Motor Company of Cortland, New York, manufactured a line of 6-ton 6x6 trucks for the military, including 1,116 bridge-erector chassis from July 1942 to March 1944. Seen here is Brockway 6-ton 6x6 chassis 512599, destined to be a bridge-erector truck. Faintly visible between the side rails of the chassis frame to the rear of the cab are two massive air cylinders that were exclusive to bridge-erector trucks. The front bumper features a rectangular opening for the cable of a front-mounted winch, a feature on some bridge-erector trucks; the winch has not been installed. (Mack Museum)

A Brockway 6-ton 6x6 bridge-erector truck, U.S.A. number 512464, is loaded with Class 50 tread plates and other bridging equipment. To the rear of the fuel tank is a large toolbox for carrying tools and equipment particular to the bridge-erecting tasks. Collapsed pneumatic floats are at the rear of the body. (US Army Engineer School History Office)

A load of steel tread plates is in the body of this 6-ton 6x6 bridge-erector truck. Each of the twin booms was operated by two hydraulic cylinders: a long one above and a shorter one below, underneath the pivot point of the boom. These hydraulic cylinders derived their power from the power take-off on the transmission. (US Army Engineer School History Office)

As seen on U.S.A. registration number 512237, a Brockway 6-ton 6x6 bridge-erector truck produced in 1942, the body had a low wall on each side, extending from the front of the body to above the space between the tandem tires. A pioneer tool rack is on the wall seen here. Also on this side were the spare-tire carrier and the tailpipe. (NARA)

Both the Daybrook Hydraulic Corporation and the Heil Company produced and installed the hydraulic booms for 6-ton 6x6 bridge-erector trucks. Here, such trucks are assembled at Daybrook's factory in Bowling Green, Ohio, and are in various stages of being fitted with the booms. (Daybrook Hydraulic)

Top left: A Brockway 6-ton 6x6 bridge-erector truck is shown in a February 1944 photograph, ready for shipment to have its body and hoists installed. For shipping purposes, the spare tire has been stowed on several wooden slats on the rear of the chassis frame. **Top right:** The same Brockway bridge-erector truck is seen from the left side with the tops of the compressed air cylinders faintly visible to the rear of the cab. These cylinders were used to inflate pontons. **Above right:** The manner in which the tailpipe was routed above and to the rear of the spare-tire carrier on the right side of the chassis is clearly portrayed. The small box above the muffler housed two ponton-inflation cocks, supplied with compressed air by the two compressed air tanks between the rails of the chassis frame. **Above left:** At 374 inches in length, the chassis of the 6-ton 6x6 bridge-erector was 88 inches longer than that of the 6-ton 6x6 prime movers. On the rear corner of the cab and the center of the cab are brackets for attaching the stanchions of a ring mount. (TACOM LCMC History Office, all)

Left: The front-mounted twin-drum winch of a Brockway 6-ton 6x6 bridge-erector truck is visible. The cathead was on the left side of the winch. The aperture in the bumper for playing out the winch cable to the front of the vehicle was offset to the right of center. Cable rollers are faintly visible inside the aperture. Heavy-duty towing rings below the bumper were part of the bridge-erector kit. **Right:** On the rear of the chassis of the Brockway 6-ton 6x6 bridge-erector truck are two bumperettes with a pintle hook between them. To the left of the pintle hook is an electrical receptacle for powering trailer lights. Below the left rear of the chassis frame is a tail-light assembly on a bracket. (TACOM LCMC History Office, both)

In this overhead view of open-cab Brockway 6-ton 6x6 bridge-erector truck chassis, Ordnance number 945, dated 24 February 1944, the twin air tanks to the rear of the cab, for inflating pontons, are highly visible. The inflation system worked separately from the air-brake system. (TACOM LCMC History Office)

The interior of the cab of Brockway 6-ton 6x6 bridge-erector truck chassis, Ordnance number 945, is seen from the right side. It was similar in appearance to the open-cab versions of the 6-ton 6x6 prime movers. Controls for the bridge-erector booms were to the rear of the cab. (TACOM LCMC History Office)

Three Purolator filters are arrayed along the left side of a Hercules Model JXD engine in Brockway 6-ton 6x6 bridge-erector truck Ordnance number 945. The two horns in the 6-ton 6x6 trucks sometimes were of polished metal, but in this case they appear to have been painted black. (TACOM LCMC History Office)

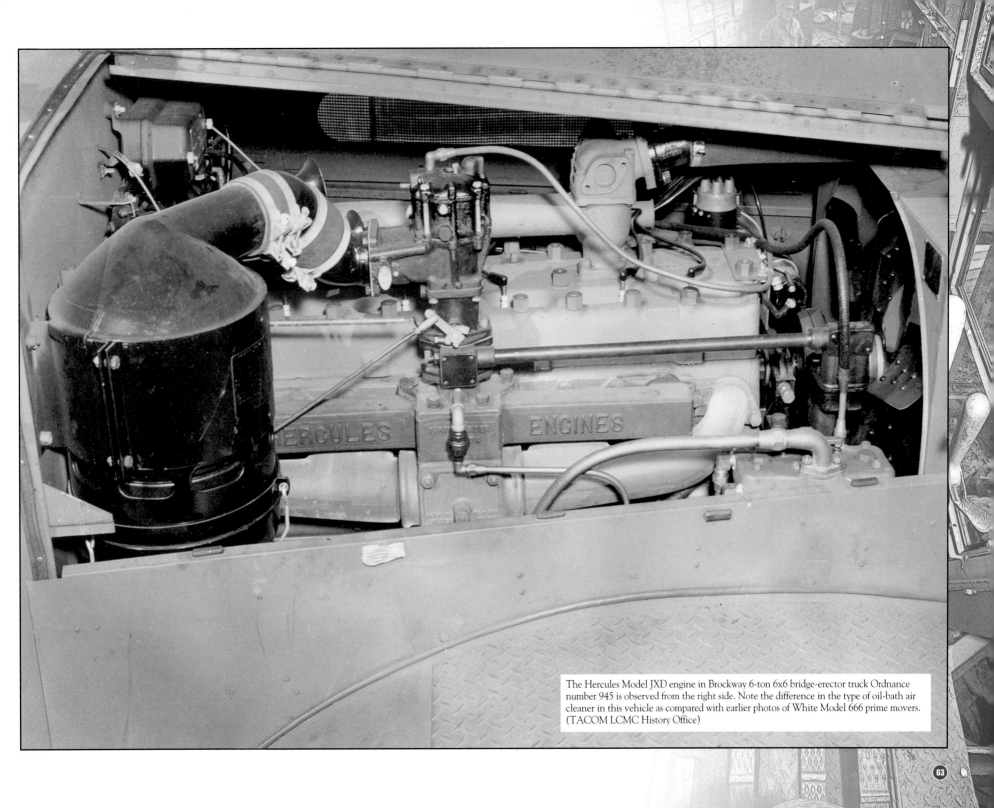

The Hercules Model JXD engine in Brockway 6-ton 6x6 bridge-erector truck Ordnance number 945 is observed from the right side. Note the difference in the type of oil-bath air cleaner in this vehicle as compared with earlier photos of White Model 666 prime movers. (TACOM LCMC History Office)

Like Brockway, White Motor Company also produced a chassis for a 6-ton 6x6 bridge-erector truck. In external appearance it was essentially the same as the Brockway version. The dual-drum front winch has been installed on this example, photographed in November 1943, and the front bumper has the rectangular opening for the winch cable. (ATHS)

Left: A clear view of the winch, the top of the hood, the cab seats and universal rifle brackets, and the rear of the chassis is available in this overhead shot of a 6-ton 6x6 bridge-erector truck. On the center panel of the hood is a rest for the windshield when it was lowered. To each side of the rest is a retaining hook. (Bryce Sunderlin collection)

Right: A Brockway 6-ton 6x6 chassis destined for a bridge-erector truck is on the company's Cortland, New York, assembly line, completed or nearly so. The soft top has been installed, but the winch cables and the bridge-erector body, booms, and equipment will be added later. (Mack Museum)

A Brockway 6-ton 6x6 bridge-erector truck is seen from the rear with the booms tilted forward in the travel position. To the upper right of the right bumperette is the opening for directing a winch cable to the rear of the truck. Prominent above and below the rectangular opening are horizontal cable rollers. (US Army Engineer School History Office)

A Brockway 6-ton 6x6 bridge-erector truck, U.S.A. number 512109, of the 23rd Armored Engineer Battalion, 3rd Armored Division, is fully loaded with equipment at a base at Fonthill Gifford, England, on 19 February 1944. In addition to the two collapsed pontons packed on the rear of the body, two complete trestles were stowed on the truck body. The nickname "EASTON" is stenciled on the top of the driver's door. (NARA)

A U.S. Army combat engineer of the 2nd Armored Division is standing at the controls of the booms of a Brockway 6-ton 6x6 bridge-erector truck, U.S.A. number 512039, during the practice launching of an inflatable ponton for an engineer bridge on the Maas River in Holland in early 1945. "ELOUISE" is painted on the driver's door, and liquid containers are secured on the running board and to the side of the truck body. A rack for machine-gun ammunition is above the cab next to the ring mount. (US Army Engineer School History Office)

Just as bridge-erector trucks were instrumental in constructing ponton bridges, they also were used to disassemble the bridges when no longer necessary, such as this Brockway 6-ton 6x6 bridge-erector truck of the 23rd Armored Engineer Battalion, 3rd Armored Division, in Wiltshire, England, in March 1944. An engineer is "riding the trestle" as the truck's hoists lift it. (NARA)

At a creek crossing, Dodge weapons carriers pass over a treadway bridge. Flanking each side of the bridge is a 6-ton 6x6 bridge-erector truck. The open-cab truck on the right is U.S.A. number 520080. The U.S.A. number of the closed-cab truck on the left is not entirely legible, but the first three digits of the six-digit number are 512. (NARA)

At the Ognon River in Voray, France, a Jeep gingerly comes alongside a 6-ton 6x6 bridge-erector truck positioned on a section of treadway bridge. The truck was being employed by U.S. Army Engineers to build a temporary bridge near a highway bridge that the retreating Germans had destroyed. (NARA)

Personnel of the 13th Engineer Battalion of the 2nd Free French Armored Division stand by as a few men on a 6-ton 6x6 bridge-erector truck maneuver a trestle section into place on the approach to a ponton bridge over the Moselle River on 10 October 1944. Clear views are available of the long upper hydraulic cylinder and the short bottom cylinder of the boom mechanism. (NARA)

Here, a bridge erector's hydraulic hoist is lifting a bridge section and placing it within reach of the Quickway crane, which appears to be mounted on a Coleman 4x4 chassis. (U.S. Army Engineer School History Office)

Members of Company A, 63rd Engineer Battalion, observe as a box crib is being lowered into the water by a 6-ton 6x6 bridge-erector truck during construction of a treadway bridge in eastern France on 8 December 1944. The crib would form a support for a section of trestle that would form an approach for the main part of a ponton bridge. (NARA)

At the same bridging site as seen in the preceding photograph, the box crib is in place in a narrow side channel of the river, and a 6-ton 6x6 bridge-erector truck is lowering a treadway section over the channel and crib to the narrow spit of land in the background. (NARA)

Negotiating a Bailey bridge in the mountains near Porretta, Italy, is a 6-ton 6x6 bridge-erector truck. This vehicle had the version of the dual-drum winch that was mounted high, rendering unnecessary an opening for the winch cable in the bumper. The personnel on the body of the truck apparently were not U.S. Army, and may have been Italians or Brazilians, as elements of the Brazilian army were occupying the front lines near this sector. (NARA)

Top: This 6-ton 6x6 bridge-erector truck U.S.A. number 0077194, has been put to interesting use at an Army Engineers base at Ubach, Germany, on 16 February 1945: its booms are holding up a 20-ton trailer on its side so that mechanics can work on its underside without lying in the mud. (NARA) **Above left:** A 6-ton 6x6 bridge-erector truck towing a trailer with a tractor-bulldozer with an armored cab crosses the Niers Canal near Viersen, Germany, en route to the Rhine River on 2 March 1945. This equipment was assigned to the 1029th Engineer Treadway Bridge Company, serving with the Ninth U.S. Army. (US Army Engineer School History Office) **Above right:** In the foreground of this view of an engineer depot along the Rhine River near Remagen, Germany, is a 6-ton 6x6 bridge-erector truck with the booms stored in the forward-pointing position. In the background is the famous Ludendorff Bridge, popularly known as the Remagen Bridge. (NARA)

Bridge-erector trucks assigned to an engineer unit with the 4th Armored Division stand by in a village in the Nancy Sector on 11 November 1944, awaiting the word to advance when required to a river crossing. On the closest truck, two deflated pontons are on the rear of the body, and a chain is draped over the cross-shaft between the two booms. (NARA)

Inflated pontons destined for a bridging operation between the cities of Winden and Kreuzau on the Roer River in Germany on 26 February 1945 are loaded on a 6-ton 6x6 bridge-erector truck assigned to the 994th Engineer Treadway Bridge Company, First Army. The pontons are lashed to the booms and to the forward part of the truck body, and the booms are deployed to the rear to help support the weight of the pontons. A helpful GI rides the front of the top ponton. (NARA)

A restored 6-ton 6x6 bridge-erector truck has been given replica markings for a vehicle attached to Company C, 55th Armored Engineer Battalion, 10th Armored Division in the European Theater of Operations. This open-cab truck is fitted with a ring mount. Loaded vertically in the body are treadway sections with steel mesh road surfaces. (John Blackman)

On each side of the rear of the bridge-erector truck, above the tandem wheels, is a long hydraulic cylinder; these operated the boom. The two shorter cylinders to the rear of the rear tires served as boosters when the boom was extended. Loaded in the rear of the body is a deflated ponton. (John Blackman)

Replica markings for Company C, 22nd Armored Engineer Battalion, 5th Armored Division are on the bumper of this 6-ton 6x6. Stenciled on the battery box door is replica shipping information of the type applied to vehicles by the U.S. Port Agency, War Shipping Administration, New York.

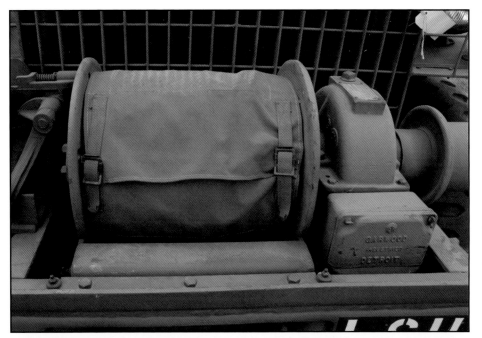

Top left: On the front of the truck is a 25,000-pound capacity PTO-driven winch. **Top right:** In addition to the center drum, the bridge erector winch is equipped with an independently engaged small drum on the passenger's side. **Above left:** The blackout driving light beamed through a special round recess in the brush guard. To the left of the blackout driving light are a blackout marker light and the headlight; to the extreme right is a non-original turning signal. **Above right:** These trucks used conventional split-ring Budd wheels rather than combat wheels.

Top left: The hoist operator had a small platform to use as a work station. Here, the platform is folded up into the stowed position, partially obscuring the instruction plate. **Top right:** The steel cab doors have a very utilitarian design. **Above left:** The large dial at left is the tachometer; the large dial at right is the speedometer. **Above right:** The safety latch on the side of the transmission gearshift lever prevented unintentional shifting into reverse. Nearest the shift pattern data plate are the dual pneumatic windshield wiper controls. To the left of the wiper controls are, from top to bottom, the blackout driving light switch, the choke control and the panel light switch.

Top left: The clutch pedal is on the left of the floorboard near the dimmer switch. The brake pedal is to the right of the steering column. Above the accelerator pedal the starter control button is visible. **Top right:** On the left side of this photo is the gear selector for the four-speed manual transmission. At right are, from top to bottom, the parking brake control, the winch control lever, the transfer case shift lever and the front axle engagement lever.

Above left: Two steel straps hold the eighty-gallon gas tank in place. The filler neck extended forward to provide enough clearance to refuel the truck from five-gallon cans. **Above right:** Note the large stowage compartment mounted just behind the fuel tank. In addition to the normal tire chains, jack, tire tools and basic hand tools, the bridge erector trucks carried additional tools needed for servicing the bed.

Top left: Two single acting hydraulic cylinders operate each lift arm. This is the upper cylinder on the driver's side. **Top right:** Heavily reinforced pivot points support the lift arms where they join the truck bed. **Above left:** A second hydraulic cylinder is mounted beneath the bed on each side. A cast steel link connects the ram of this cylinder to the lift arm. **Above right:** The flexible and somewhat fragile hydraulic lines are protected by the pivot point structure.

Top left: The lower cylinders take over the operation of the lift arms about halfway through their extension. **Top right:** Details of the mounting for the right boom of a 6-ton 6x6 bridge-erector truck are shown. The reflectors and tail lights are recessed behind round openings in the rear cross member of the body. Below the reflector is a tie-down ring. To the left is part of the fairlead assembly. **Above left:** The long hydraulic cylinder on the right side of the body is viewed from the left corner of a 6-ton 6x6 bridge-erector truck. The hydraulic cylinders that operated the booms were served by a hydraulic pump that was powered by the engine via a power take-off. The booms had a maximum capacity of 8,000 pounds.

Above right: Connecting the tops of the dual booms of the 6-ton 6x6 bridge-erector truck is a truss constructed of metal tubes and steel strapping in the form of X-braces. On each upper corner of the bulkhead at the front of the body is a spotlight for illuminating the work area during night operations not under blackout conditions.

Crane Truck

Brockway produced a chassis for mounting a Quick-Way crane with a shovel. It was based on the 6-ton 6x6 chassis. Brockway built a total of 1,224 crane chassis during a production run lasting from July 1943 through June 1945. A test chassis without the crane installed is shown in a 16 June 1944 photograph. The vehicle featured an open half-cab to provide clearance to its side for the crane boom in the travel position. The spare tire was to the rear of the cab. (TACOM LCMC History Office)

A quantity of Brockway 6-ton 6x6 crane chassis are lined up prior to the installation of the cranes. Like the 6-ton 6x6 bridge-erector truck, the crane chassis had a double-drum front winch rated at a capacity of 25,000 pounds. Faintly visible on the hoods of the first two trucks to the front right are the U.S. Army registration numbers, 0050969 and 0050970. (Grumman Memorial Park via Leo Polaski)

A Brockway 6-ton 6x6 crane truck with the crane installed has been secured to a railroad flatcar for shipment. Sealant has been applied to the body joints of the crane cab, which is rotated to the rear, and the crane boom has been disassembled; the lower half of the crane is secured to the travel lock, and the upper part of the crane is stowed above the dual tandem tires. (Grumman Memorial Park via Leo Polaski)

A Brockway 6-ton 6x6 crane truck is partially disassembled and packed for shipment. For packing, the wheels and tires have been removed from the axles and are stowed on a platform, with the top of the crane travel lock protruding through the platform. The crane boom has been removed and secured to the side of the vehicle. Fragile parts have been removed and stored in boxes. (Grumman Memorial Park via Leo Polaski)

The Quick-Way Truck Shovel Company of Denver, Colorado, was a major supplier of cranes for the Brockway 6-ton 6x6 crane trucks, an example of which is shown in this photograph. The type employed was the Quick-Way Model E crane. The roof of the Quick-Way crane cab had noticeably different contours than the Osgood crane cab. (US Army Engineer School History Office)

U.S. Army registration number 0091660 is stenciled on the crane cab and on the hood of this Brockway crane truck. Caution stripes, probably black and yellow, are painted on the lower rear of the crane cab. A hook is on the end of the crane cable, but these cranes were equipped with a variety of accessories, such as a dragline bucket, a clamshell bucket, a pile-driver, a dipper bucket, and a crane hook. (US Army Engineer School History Office)

Left: Brockway 6-ton 6x6 crane trucks frequently operated with engineer bridging units. Here, a crane truck is preparing to lift an inflated ponton from a 6x6 cargo truck. Markings for the 996th Engineer Treadway Bridge Company are on the bumper of the crane truck. The Quick-Way crane had a maximum safe lifting capacity of 13,000 pounds, depending on the angle and reach of the crane. (US Army Engineer School History Office)

Right: A Brockway crane truck with a Quick-Way crane unloads crated supplies from a 6x6 cargo truck at a depot in France on 6 September 1944. The U.S.A. number, 0051146, is marked in white on the hood of the crane truck, and large recognition stars are painted on the crane cab. "ASCZ," for Advance Section, Communications Zone. (NARA)

Members of Company B, 1553rd Engineer Heavy Pontoon Battalion, Seventh Army, employ a Brockway crane truck to hoist a rigid steel pontoon weighing 4,400 pounds from a trailer somewhere in France on 24 September 1944. The 1553rd was an African-American unit. (NARA)

U.S. Army Engineers use a Brockway crane truck to lift treadway sections for a ponton bridge across the Moselle River at Neuviller-sur-Moselle in France on 12 September 1944. On the front of the crane cab is a rack holding three German Jerrycans. Note the raised window on the front top of the crane cab. (NARA)

A Brockway crane truck, U.S. Army registration number 0051001, is about to drive off a five-ponton raft. This was part of a river-crossing exercise in preparation for future operations in Germany. An unidentified, girder-type assembly is lashed to the crane travel lock. (NARA)

At the close of World War II in Europe, on 17 May 1945 a column of trucks, including a Brockway crane truck at the front, is being prepared to leave Depot E5084 in Villeneuve, France, with an intended destination of the China-Burma-India (CBI) Theater. Written on the side of the crane cab is "C.B.I. HERE WE COME." The truck's U.S.A. number is 0051158. (Bryce Sunderlin collection)

In another scene from Depot E5084 at Villeneuve, France, on 17 May 1945, a Brockway crane truck prepares to depart. Hitched to the truck is a trailer is a Timpte Model QW-T-8 trailer, a standard trailer for transporting accessories for the crane truck. Visible on the trailer are a clamshell shovel and a shovel boom. (NARA)

A Quick-Way crane on a Brockway crane truck, U.S.A. number 0079236, deposits lumber on a railroad flatcar at Port Legaspi, Luzon, Philippine Islands, on 1 May 1945. The lumber was destined for railroad bridge construction. The light-colored radiator for the crane engine is visible through the door in the crane cab. (Bryce Sunderlin collection)

Among the heavy equipment being loaded on several landing craft, tank (LCTs) at the Po-Hang LST Basin near Yeongdeok, Republic of Korea, during the opening days of the Korean War, on 28 July 1950, is a Brockway crane truck, Note the folded-down windshield and the narrow bow at the rear of the cab. U.S. Army registration number WE925001 is on the hood. (NARA)

Brockway crane truck U.S.A. number 0086513 is equipped with a shovel boom, a dipper stick, and a dipper. Among the equipment lying on the ground are a pile-driver assembly, a crane boom, a dragline bucket, a clamshell scoop and a crane hook. "WORK SAFELY" is painted on the shovel boom. (US Army Engineer School History Office)

Brockway crane truck U.S.A. number 0086513 is shown in another photo, again with a shovel boom, dipper stick, and dipper installed. The dipper bucket is resting on the travel lock. The dipper was used largely for excavating work. The depth of the chassis frame from the rear of the cab to the rear wheels is shown to good advantage. (US Army Engineer School History Office)

Photographed at a War and Peace military equipment show in the U.K. was this nicely restored Brockway 6-ton 6x6 crane truck with a Quick-Way crane. Here, it is hoisting a large crate into a 6-ton stake trailer connected to a Chevrolet 1½-ton 4x4 Truck Tractor. The top windshield of the crane cab is often seen open in vintage photos of crane trucks, for ventilation and better visibility for the crane operator. (John Blackman)

A sliding door was located on the left rear side of the crane cab, shown closed here. It provides access and ventilation to the radiator and engine that powers the crane. This truck had the later-type walking-beam tandem suspension instead of leaf springs. (John Blackman)

Left: The Quick-Way crane as mounted on the Brockway 6-ton 6x6 crane truck was rated to lift up to 12,000 pounds when the crane was operating to the front or the rear: that is, along the truck's centerline. When operating to the sides, the crane could lift up to 13,000 pounds when outriggers were installed. As the elevation angle of the crane decreased, the lifting ca- pacity declined. **Right:** A close-up view of the Quick-Way crane cab shows the arrangement of cable sheaves above and below the open front door. Between the legs of the foot, or base, of the crane is an assembly containing several wheels on a frame; this is the self-aligning fairlead for the dragline bucket. (John Blackman, both)

Left: An operator is at the controls of the Quick-Way crane. The booth has a sliding door with padlock hasps installed. Above the door is a running light. The operator's seat is a simple tractor type on a spring-steel mount. To the front of his foot is the hoist brake pedal.
Right: This is a driver's view of the controls of the Quick-Way crane as mounted on a Brockway 6-ton 6x6 crane truck. The levers are, left to right, the engine clutch lever, the push and pull swing lever, the push-raise boom and pull-haul back drum lever, and the pull-hoist drum lever. Below the lower right corner of the windshield is the hydraulic fluid supply tank. (John Blackman, both)

GANTRY FRAME

DRUM GUARDS

REAR HOIST DRUM

BOOM DRUM LOCK

BOOM BRAKE PEDAL

HAUL BACK BRAKE PEDAL

HOIST BRAKE PEDAL

ENGINE CLUTCH LEVER
PUSH - SWING LEFT
PULL - SWING RIGHT

PUSH - RAISE BOOM
PULL - HAUL BACK DRUM
PULL - HOIST DRUM

QW/09 INT. MOTOR

GEAR GUARD

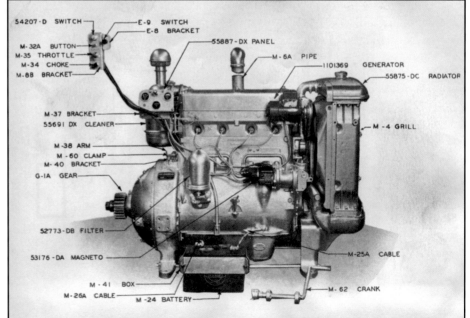

54207-D SWITCH
E-9 SWITCH
E-8 BRACKET
M-32A BUTTON
M-35 THROTTLE
M-34 CHOKE
M-8B BRACKET
55887-DX PANEL
M-6A PIPE
1101369 GENERATOR
55875-DC RADIATOR
M-37 BRACKET
55691 DX CLEANER
M-4 GRILL
M-38 ARM
M-60 CLAMP
M-40 BRACKET
G-1A GEAR
52773-DB FILTER
53176-DA MAGNETO
M-25A CABLE
M-41 BOX
M-26A CABLE
M-24 BATTERY
M-62 CRANK

Top left: In the rear of the crane cab is the engine that powers the crane. The specified engine for the Quick-Way crane and shovel as mounted on the Brockway crane truck was the International Harvester Model Q-W/U-9, a 4-cylinder, 35-horsepower gasoline engine. In the foreground is the rear hoist drum. **Top right:** The Quick-Way crane cab is shown from the right side with the cab enclosure removed. Toward the right is the gantry frame and the operator's seat and controls, to the rear of which are the cable drums and guards and the motor. **Above left:** In a view of a Quick-Way crane cab with the enclosure removed, the Quick-Way crane motor is in the foreground, with the fan and the radiator to the left. To the right of the engine is the gear guard. **Above right:** The Quick-Way Q-W/U-9 motor is seen from the right side. To the right is the radiator, and in the lower foreground are the battery and the motor crank. To the upper right are the engine control switches.

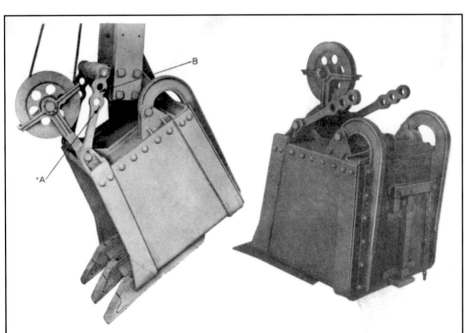

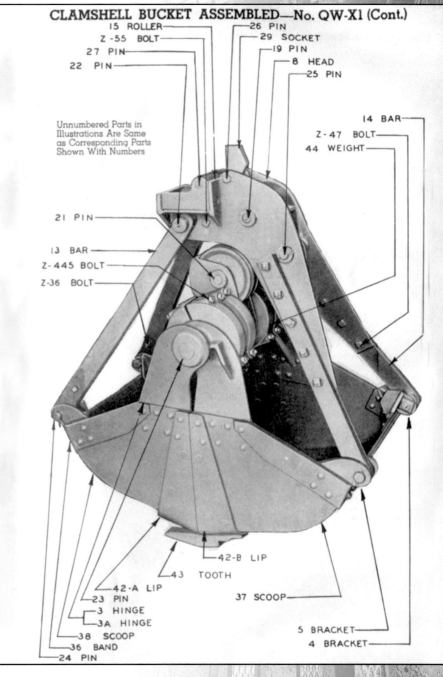

CLAMSHELL BUCKET ASSEMBLED—No. QW-X1 (Cont.)

15 ROLLER
Z -55 BOLT
27 PIN
22 PIN

26 PIN
29 SOCKET
19 PIN
8 HEAD
25 PIN

14 BAR
Z - 47 BOLT
44 WEIGHT

Unnumbered Parts in Illustrations Are Same as Corresponding Parts Shown With Numbers

21 PIN

13 BAR
Z - 445 BOLT
Z-36 BOLT

42-B LIP
43 TOOTH
42-A LIP
23 PIN
3 HINGE
3A HINGE
38 SCOOP
36 BAND
24 PIN
37 SCOOP
5 BRACKET
4 BRACKET

Top left: A photograph from a Quick-Way crane manual shows the dragline shovel that was included with each crane assembly. The dragline shovel was suspended from the crane boom, and a dragline attached to the shovel pulled it along the surface to scoop up material. **Above left:** A dipper bucket also was part of the Quick-Way crane accessories. The dipper was a type of excavating shovel that was suspended from a rigid boom called a dipper stick. **Right:** A clamshell bucket as used with the Quick-Way crane is depicted in an image from a technical manual. The clamshell bucket comprised two hinged sections that opened up to scoop up material, and then closed shut until the material was to be released.

Tanker Truck

The White Motor Company built a number of gasoline tank trucks based on its Model 666 prime-mover chassis, including, reportedly, 25 examples from May to August 1943. The tanks on these trucks had a capacity of 2,000 gallons, divided into four self-sealing cells, and the Butler Manufacturing Corporation supplied the tanks. Testing of at least one example of the White 6-ton 6x6 2,000-gallon gasoline tanker was underway by 26 December 1941, when this photograph of such a vehicle was taken. (NARA)

512 GAL. 498 GAL. 500 GAL. 490 GAL.

In a right-rear view of a White 2,000-gallon gasoline tanker, bows are installed for supporting a tarpaulin for camouflaging the vehicle as a cargo truck. **Inset:** With the doors at the rear of the White 2,000-gallon gasoline tanker open, the fuel hoses, valve controls, and, toward the bottom of the compartment, the spigots, one for each cell of the fuel tank, are visible. (NARA, both)

This gasoline tanker truck, U.S.A. number 0079191, was based on a 1943 Brockway C666 chassis originally intended to mount a Quick-Way crane. Since fuel trucks were high-priority targets, production 6-ton 6x6 gas tanker trucks and trailers were equipped with removable bows and tarpaulins, in order to camouflage the vehicles as cargo vehicles. (US Army Quartermaster Museum) **Inset:** In a 26 December 1944 photograph, Brockway 2,500-gallon gasoline tanker U.S. Army registration number 0079191 and the tanker trailer hitched to it are fitted with tarpaulins, and the open-type cab has the cover and door curtains with clear plastic windows. In the original caption to this photo by the Chief of Ordnance, Detroit, the truck is designated the "Truck, 6 ton, 6x6, Tank, Gasoline, 2500 Gallon, M34," and the trailer is designated the "Trailer, 7$\frac{1}{2}$ Ton, 4 Wheel, Tank, Gasoline 2500 Gallon, M27." (NARA)

A 6-ton 6x6 chassis with a 2,500-gallon gasoline tank, based on a Brockway chassis for a Quick-Way crane, is configured to tow a 2,500-gallon gasoline tank trailer. The front of the trailer was of a beveled design, to prevent the front corners of the tank from striking the rear of the tank on the truck during tight turns. (NARA)

The 2,500-gallon versions of the 6-ton 6x6 gasoline tanker and trailer had a squared or boxy structure below and abutting the rounded part of the sides of the tank, as seen here, whereas the 2,000-gallon tanks had an angled or beveled structure below and abutting the rounded sides of the tanks. (US Army Quartermaster Museum)

The U.S.A. number of this 6-ton 6x6 tank truck, 0051086, identifies it as based on a Brockway chassis for a Quick-Way crane, and it had a dual-drum front winch and bumper with cable opening. According to the original caption of the photo, the tank had a 2,500-gallon capacity, and the truck was towing a 2,500-gallon tank trailer. (US Army Quartermaster Museum)

Fire Truck

Brockway built 242 chassis for Model F666 Class 155 6-ton 6x6 fire/crash trucks for the U.S. Army Air Forces from March 1944 through June 1945. The chassis were fitted with Mack fire-fighting equipment. Two turrets atop the body delivered high-pressure fog foam to combat fires, and two flexible lines also were available for spraying foam. Shown here is U.S.A. number 507340. (National Museum of the United States Air Force)

Above the rear fender of the Model F666 Class 155 6-ton 6x6 fire/crash truck is a bin for stowing the hose for one of the two portable foam sprayers; the sprayer is stowed on the side of the body to the front of the bin. A fire extinguisher is on the running board, a siren is mounted on the front bumper, and a small spotlight is on the top of the left side of the windshield frame. (Jim Davis collection)

On the Brockway Model F666 Class 155 fire/crash truck, the front winch was removed, although the winch-cable aperture in the bumper remained. An A-frame with a heavy-duty clevis at the top was installed in front of the grille for purposes of towing. A ladder on each side of the body provided access to the top of the body. The vehicle could carry 1,000 pounds of water. (NARA)

In a view of Brockway Model F666 Class 5 fire/crash truck U.S. Army registration number 507686, at the rear of the body is the compartment where the centrifugal pump and its motor were located. The motors for these vehicles were produced by American LaFrance and Continental. (National Museum of the United States Air Force)

Tractor Truck

Due to overloading of the rear axles when used as a truck-tractor, the Chief of Engineers contracted with the Sterling Motor Truck Company to install a chain drive rear bogie on a 6-ton 6x6. This vehicle, designated T27E1, was the result of this contract. (Tony Gibbs collection)

As part of the adaptation of the 6-ton to truck tractor configuration, 14.00-20 tires were to be used, which further increased the loads on the axle bearings. This Sterling T27E1, as well as an alternate prototype developed by Cook Brothers, designated the T27E2, both resolved these problems, but work was terminated in August 1945 without either type being adopted as standard. (Tony Gibbs collection)

During World War II, around 112 tractors were produced based on White Model 666 chassis. They were intended to tow the 20-ton low-bed, front-loading semitrailer. For that purpose, a fifth wheel was mounted on the chassis frame. In addition heavier-duty Timken front and tandem axles and larger tires than those used on the White 666 prime mover were employed. (ATHS)

On the White 6-ton 6x6 tractor, the winch behind the cab was eliminated, and a front winch of 25,000 pounds capacity was installed. Included was a front bumper with a winch-cable opening and cable rollers. This was a single-drum, single capstan model. The fuel filler was the elbow type located on the upper front of the fuel tank. (ATHS)

The fifth wheel of the White tractor was located above the tandem wheels. Behind the cab was a free-standing locker, at the top of which was a frame that held an electrical receptacle and service and emergency air couplings. (Bryce Sunderlin collection)

Left: A frontal view of a White 666 tractor shows the upper part of the Gar Wood single-drum winch assembly. Visible inside the rectangular opening for the winch cable in the bumper are horizontal and vertical rollers. Note the ring incorporated in the left corner of the grille for the blackout headlight. **Right:** For hauling trailers, the rear of the chassis frame of the White 6-ton 6x6 tractor was equipped with a tow pintle, to the lower left of which was a receptacle for supplying electrical power to the trailer. The maker's name, Layton, is faintly visible on the fifth wheel. (Bryce Sunderlin collection, both)

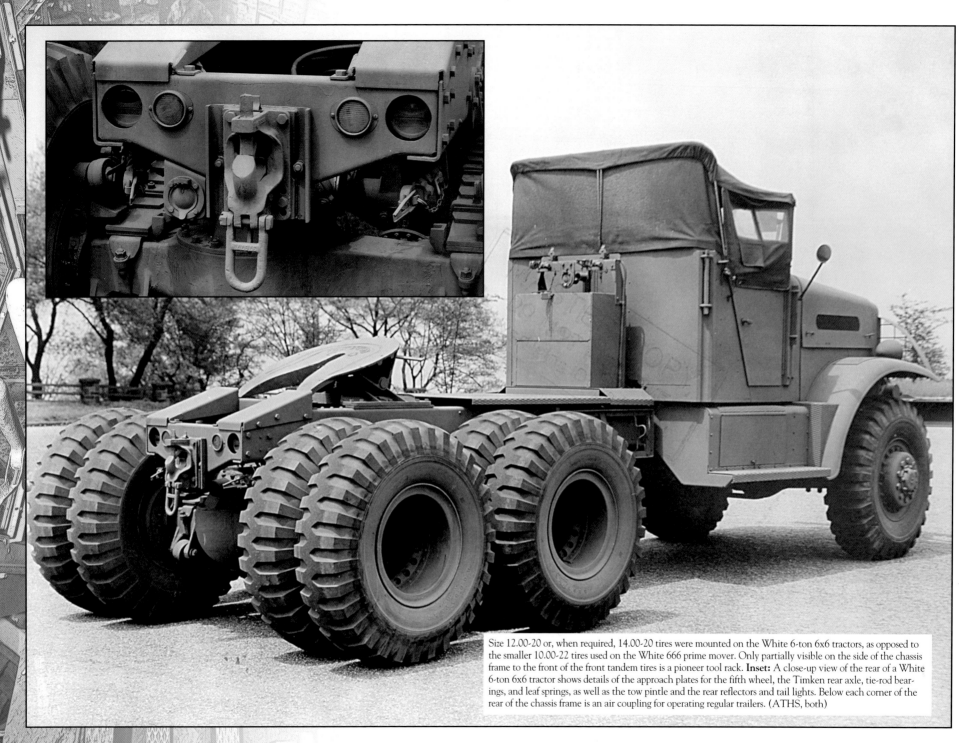

Size 12.00-20 or, when required, 14.00-20 tires were mounted on the White 6-ton 6x6 tractors, as opposed to the smaller 10.00-22 tires used on the White 666 prime mover. Only partially visible on the side of the chassis frame to the front of the front tandem tires is a pioneer tool rack. **Inset:** A close-up view of the rear of a White 6-ton 6x6 tractor shows details of the approach plates for the fifth wheel, the Timken rear axle, tie-rod bearings, and leaf springs, as well as the tow pintle and the rear reflectors and tail lights. Below each corner of the rear of the chassis frame is an air coupling for operating regular trailers. (ATHS, both)

Hitched to this White 666 tractor is a 20-ton front-loading semitrailer with a D8 tractor-dozer loaded on it. The front end of this semitrailer formed a gooseneck when hitched to the fifth wheel, and the front end could be lowered to form a ramp for boarding and dismounting vehicles. (US Army Engineer School History Office)

This fifth-wheel tractor photographed at the headquarters of the 709th Engineer Construction Battalion in Korea on 10 April 1953 is somewhat of an anomaly, as it lacks a front winch and retains a prime-mover-type front bumper and center-mounted winch. The vehicle is based on a White 666, and its U.S.A. number, 5115219, falls within the range of U.S.A. numbers for Model 666 tractors as listed in the Ordnance Depart list of Administrative and Tactical Vehicles, 1940-1945. Hitched to the tractor is a 20-ton rear-loading semitrailer. (NARA)